Secrets of the Num

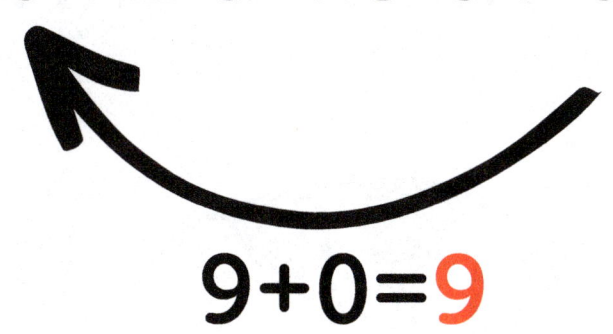

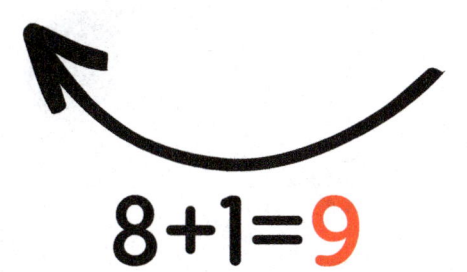

0 1 2 3 4 5 6 7 8 9

9+0=9

0 1 2 3 4 5 6 7 8 9

8+1=9

0 1 2 3 4 5 6 7 8 9

7+2=9

0 1 2 3 4 5 6 7 8 9

6+3=9

0 1 2 3 4 5 6 7 8 9

5+4=__

Color and Write

nine 9 9 9

TABLE 9 always add 1 up and down

9x1=9

9x2=18
+1 ↘ ↗ +1
9x3=27
+1 ↘ ↗ +1
9x4=36
+1 ↘ ↗ +1
9x5=45   **9x5=45=(4+5=9)**
+1 ↘ ↗ +1
9x6=54
+1 ↘ ↗ +1
9x7=63   **9x7=63=(6+3=_)**
+1 ↘ ↗ +1
9x8=72
+1 ↘ ↗ +1
9x9=81
+1 ↘ ↗ +1
9x10= 90

TABLE 9 always add 1 up and down

9x1=9

9x2=___
+1 ⟩ ⟩ +1
9x3=27
+1 ⟩ ⟩ +1
9x4=36
+1 ⟩ ⟩ +1   **9x5=45=(4+_=9)**
9x5=_ 5
+1 ⟩ ⟩ +1
9x6=54
+1 ⟩ ⟩ +1
9x7=63   **9x7=63=(6+_=9)**
+1 ⟩ ⟩ +1
9x8=72
+1 ⟩ ⟩ +1
9x9=___
+1 ⟩ ⟩ +1
9x10= 90

I CAN COUNT!

1 2 3
4 5 6
7 8 9

TABLE 9 the mirror image

9x1=9

9x2=18 <-> 81

9x3=27 <-> 72

9x4=36 <-> the mirror _

9x5=45 <-> the mirror __

9x6=**54 <-> 45**

9x7=63 <-> 36

9x8=72 <-> the mirror __

9x10= 90 <--> the mirorr is 09

9

TABLE 99 always add 1 up and down and put 9 in the middle 2**9**7

99x1 = 99 briefing

99x2=1**9**8

+1 ⤵ ⤴ +1

99x3=2**9**7

+1 ⤵ ⤴ +1

99x4=3**9**6

+1 ⤵ ⤴ +1

99x5=4**9**5

+1 ⤵ ⤴ +1

99x6=5**9**4

+1 ⤵ ⤴ +1

99x7=6**9**3

+1 ⤵ ⤴ +1

99x8=7**9**2

+1 ⤵ ⤴ +1

99x9=8**9**1

+1 ⤵ ⤴ +1

99x10= 9**9**0

9

TIME IS EVERYTHING

We measure our lives in days and hours, let's explore where is number 9

24 hours =1440 minutes =
(1+4+4=_)

These numbers are carrier of information

1440 minutes =86400 seconds
(8+6+4=_=_)

These numbers are carrier of information

720 minutes = 43200 seconds
(4+3+2=_)

These numbers are carrier of information

3-4 age

PRACTICE IN RECOGNISING THE NUMBER 9

Make 9 with your finger.

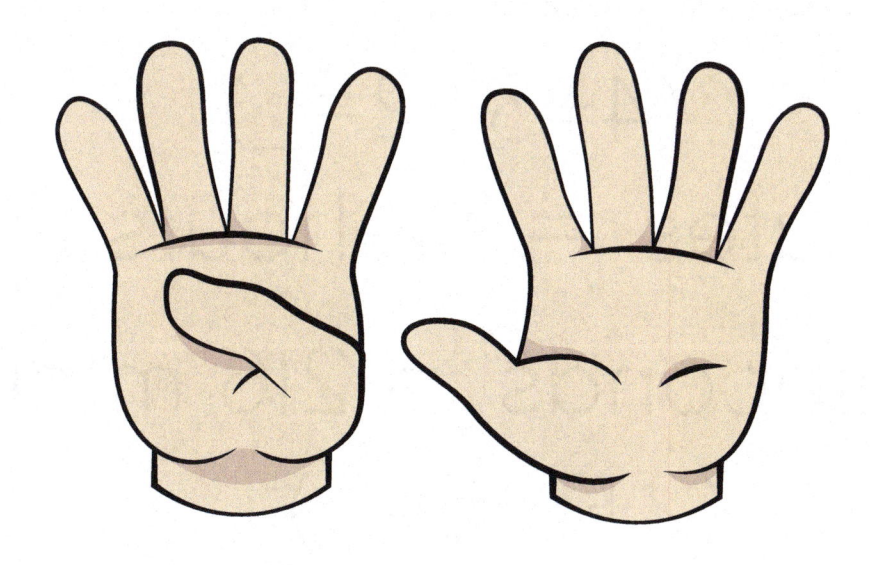

Trace over the 9 with your finger.

Count how many planets there are.

Exercices for children

Name-Surname: Class:

TIME IS EVERYTHING

Sacred Numbers Embedded in Time

25'920 seconds = 432 minutes = (4+3+2=_)

432 minutes = 7,2 hours (7+2=_)

12'960 seconds = 216 minutes (2+1+6=_)

6'480 seconds = 108 minutes (1+0+8=_)

25'920 hours = 1080 days (1+0+8=_)

These numbers are carrier of information

3-4 age

NUMBERS / PRACTICE FOR WRITING THE NUMBER 9

Write the number 9.

Exercices for children

Name-Surname: Class:

SPACE MEASUREMENT

Each number stands for meaning beyond the appearance

1 foot = 12 inches

These numbers are carrier of information

1 square foot = 144 inches

1 + 4 + 4 =
144 cubits in the holy city !

1 cubic foot = 1728 cubic inches

1 square yard = 1296 inches

12960 = main cycle for the planet

3-4 age

NUMBERS / PRACTICE FOR WRITING THE NUMBER 9

Trace over the number 9 with a pencil.

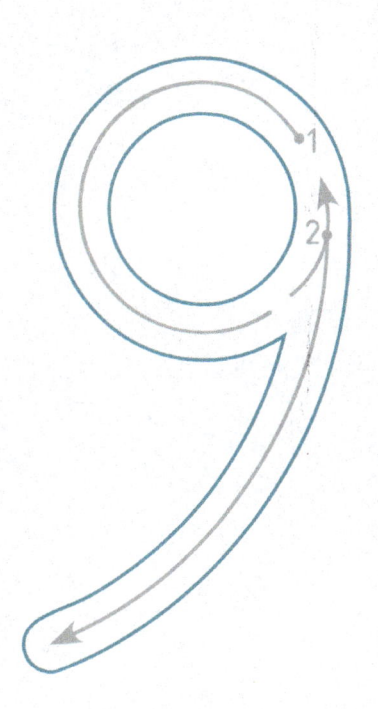

Colour 9 baloons.

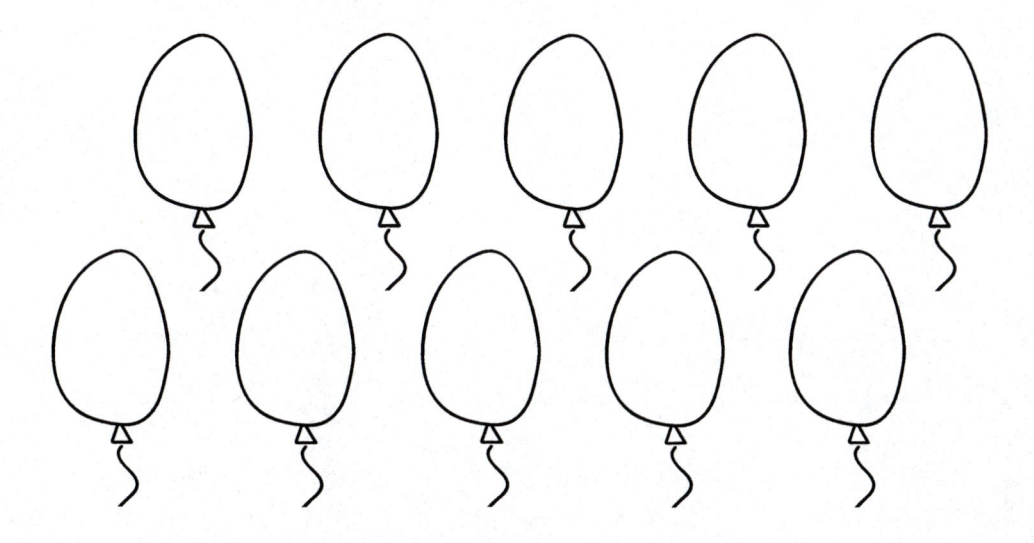

Exercices for children

Name-Surname: Class:

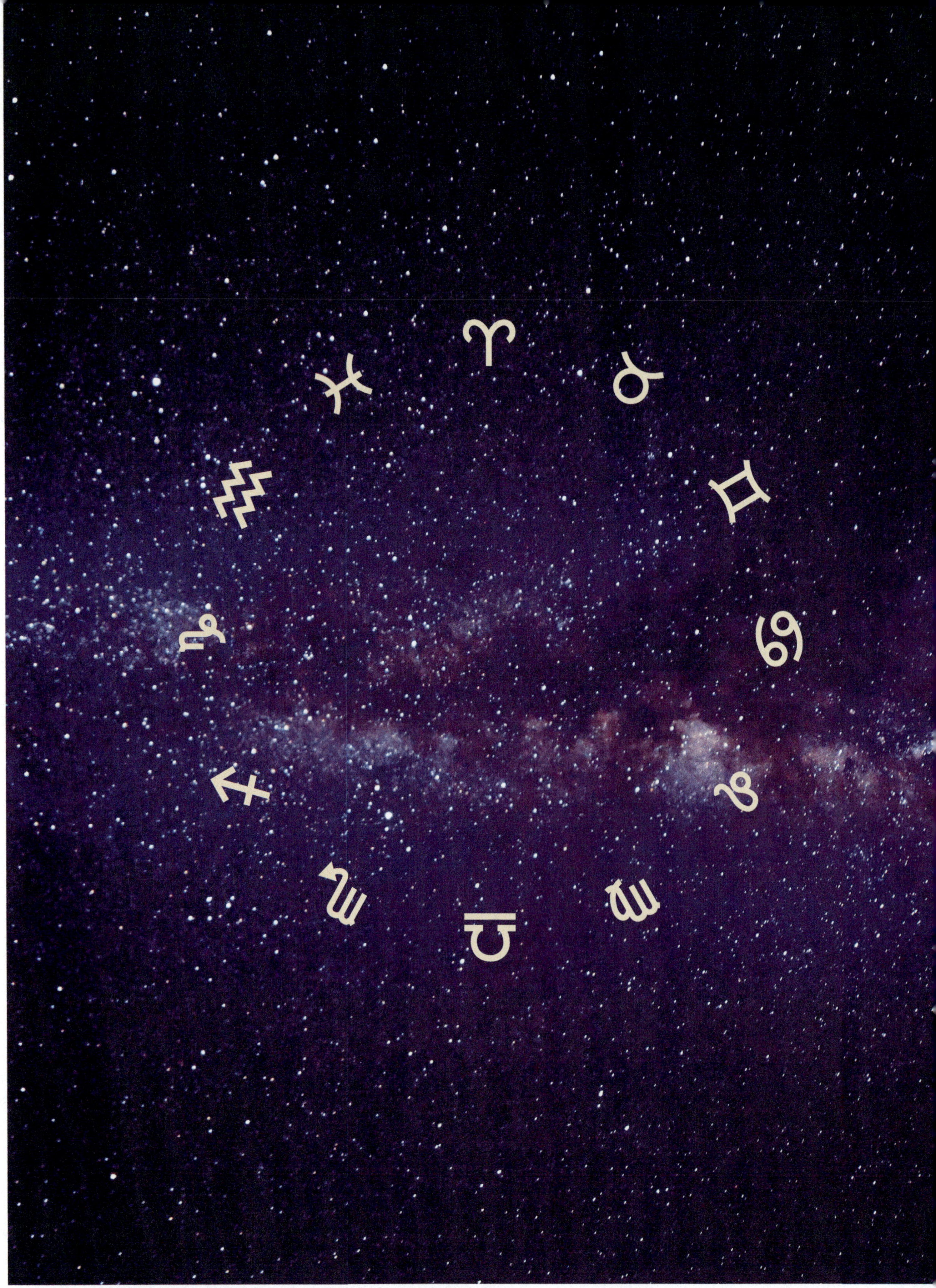

VEDIC AGES & MATH

Each number stands for meaning beyond the appearance

These numbers are carrier of information

VEDIC MATHS

144 cubits in the holy city !

12960 = main cycle for the planet

TABLE NO.

9

4 seasons = 6480 years each

Gothic cathedrals, 6480 x 4 = 25920 years = 2+5+9+_=18=1+8=

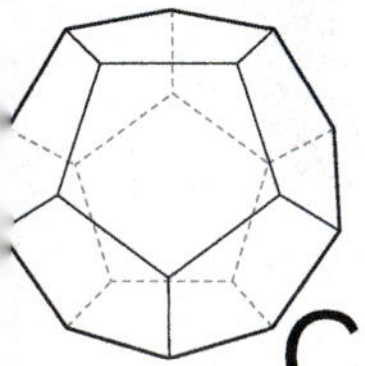

Cyclical time, 6480 years / degrees

Cyclical time, 3600 years / degrees

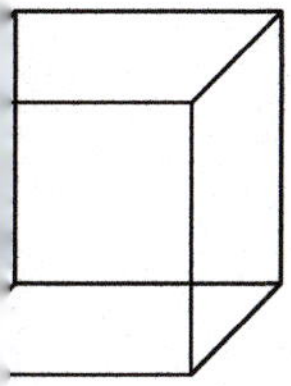

Cyclical time, 2160 years / degrees

Cyclical time, 1440 years / degrees

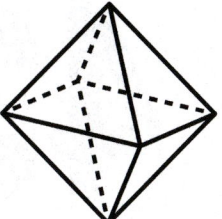

Cyclical time, 720 years / degrees

Sacred Calendar **360 days =3+6+**

(5,25 days unlucky days to reach 365!)

12'960 = main cycle for the planet

They ALL reduce to 9 !

1440

2160

8640

43200

1296

1728

36

They ALL reduce to 9 !

9

Do it, and discover the magnificience of number 9

Exercices

1440 = 1+4+_+_ =

8640 = _+_+_+_ =

43200 = _+_+_+_+_ =

1296 = 1+_+9+_ =

1728 = 1+7+_+_ =

36 = _+_ =

(9)

FOOT / SOCCER

Each number stands for meaning beyond its appeareance

Half Match is = 45 minutes
4+5=

Total Match = 90 minutes

These numbers are carrier of information

1 cubic foot = 1728 cubic inches

1 square yard = 1296 inches
12960 = main cycle for the planet

144 cubits in the holy city !

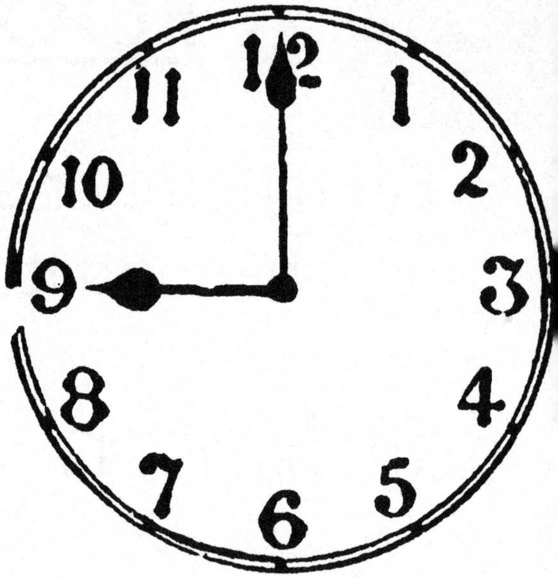

9

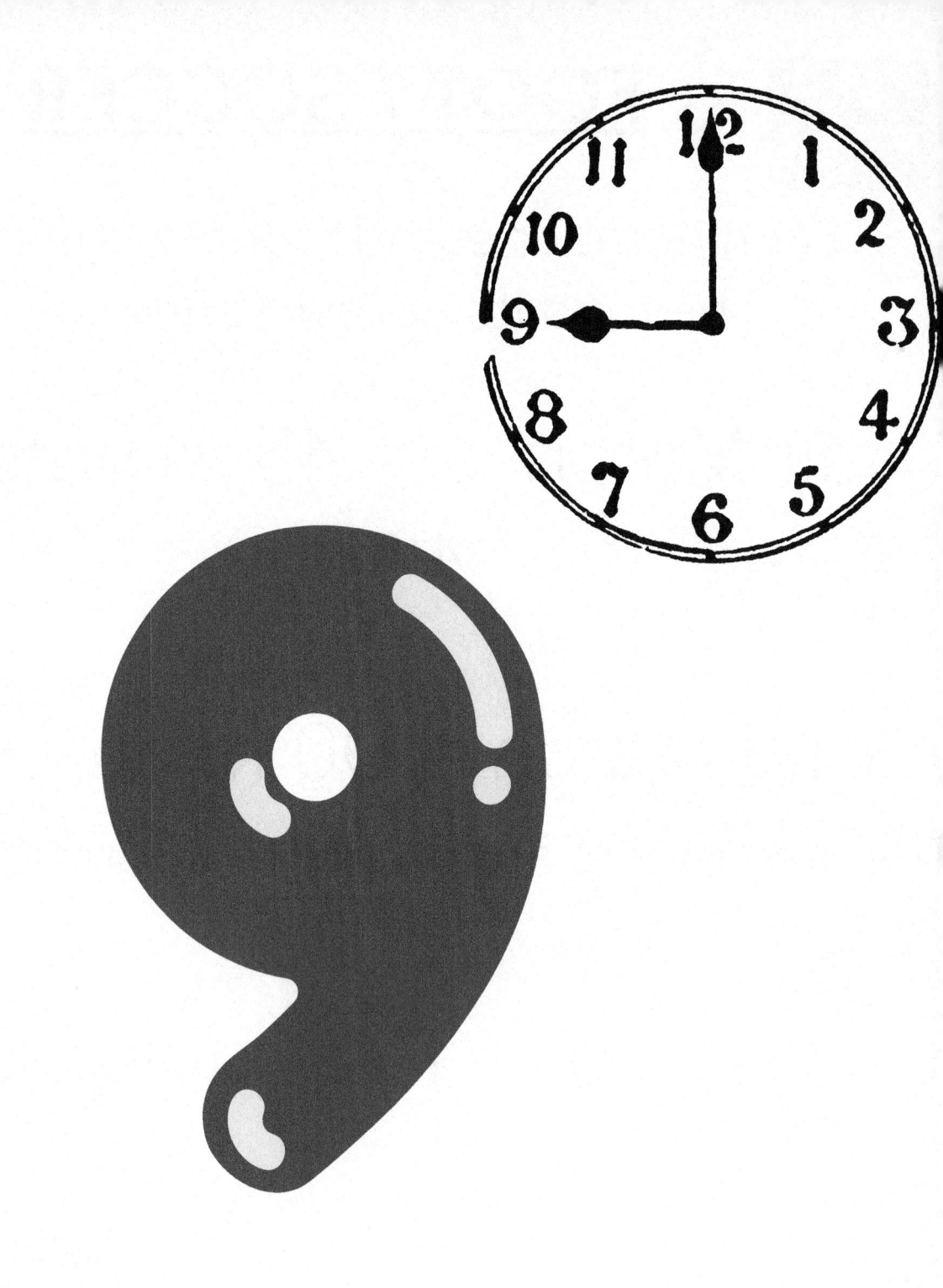

9

3,6,9

Exercices

Diameter of the Sun 864000 miles

864000 = _____ = 18 = 9

Diameter of the Earth 7920 miles

7920 = 7 + 9 + _ = 9

Diameter of the Moon: 2160 miles

2160 = _ + _ + 6 = 9

Have you ever tried to divide 9 by 2 ?

reduce it = + until impossible...

9/2 = 4,5 = 4+5 =

4+5=9
Let's divide by 2 again

4,5/2 =

2.25/2 =
Let's divide by 2 again

1.125/2 =
Let's divide by 2 again

0,5625 = 5+6+2+5 =
Let's divide by 2 again

Exercices

9

9 IS EVERYTHING

$9 \times 1 = 9$

$1+2+3+4+5+6+7+8+9 = \underline{} = 9$

9 is sustained by 3 and 6

Exercices

```
        9
        △
      /   \
     /     \
    6-------3
```

9 IS THE NUMBER of COMPLETION

$3 + 6 = \underline{}$

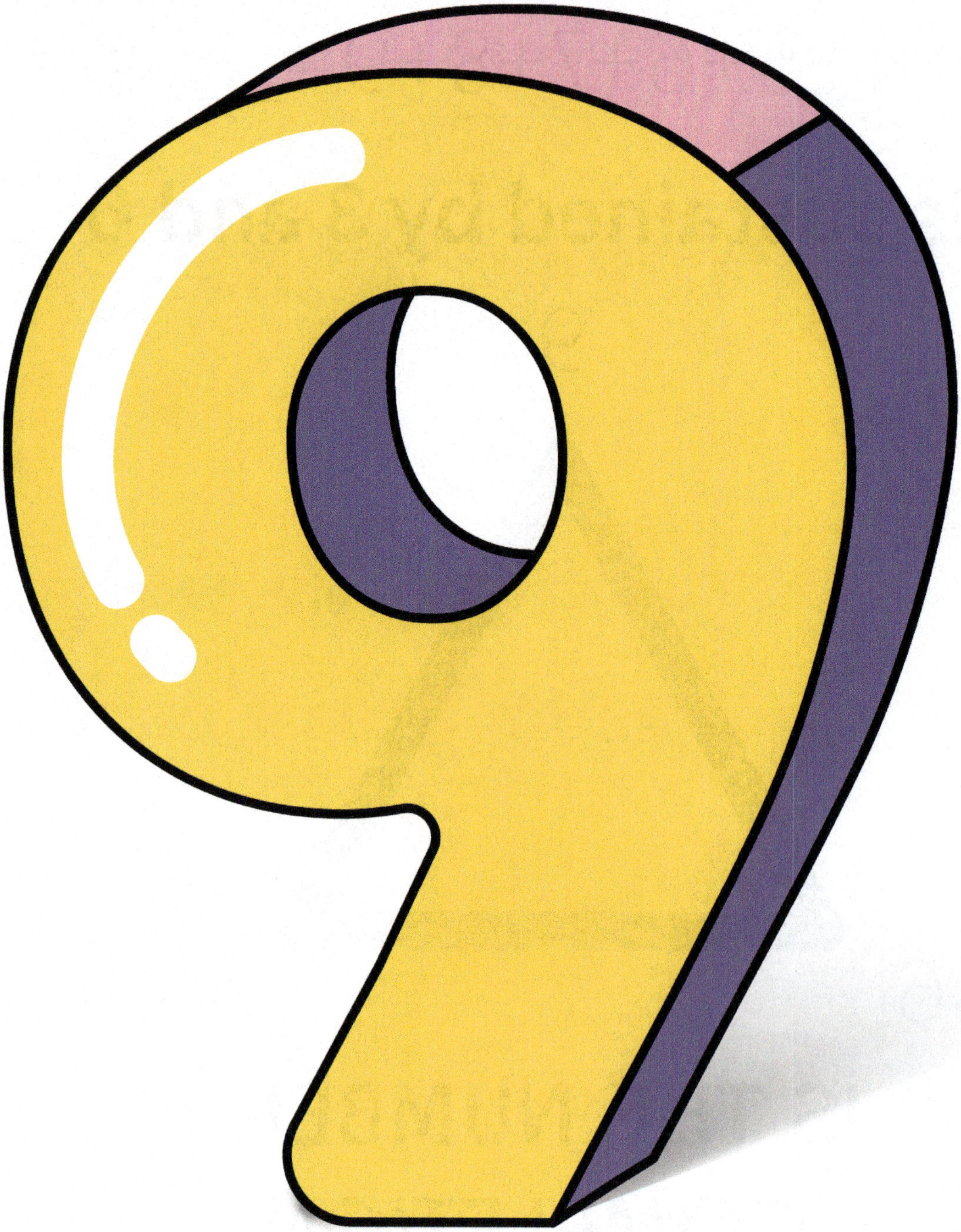

Exercices

The doubled game always end with:
1,2,4,8,7,5

The doubled game: 1+1=_

The doubled game: 2+2=_

The doubled game: 4+4=_

The doubled game: 8+8=__ = 7

The doubled game: 16+16=__ = 5

The doubled game: 32+32=__ = 10=1

The doubled game: 7+7=__ = 5

9 is free and in NO patterns

9 is absent of the doubled game

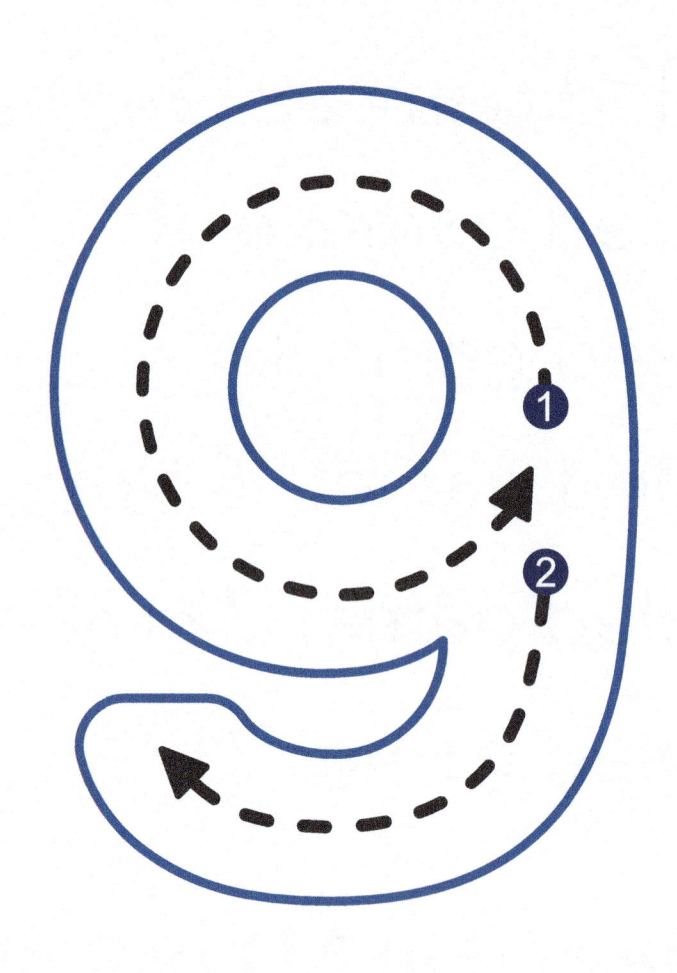

Exercices

The doubling game with 3 and 6 now !
9 is absent again !

The doubled game: 3+3=_

The doubled game: 6+6=_=1+_=

The doubled game 12+12=_=2+_=

The doubled game 48+48=_+_= 1+5=6

The doubled game 96+96=192= _+_+_=3

The doubled game 192+192=384= 3+8+4=1+5=6

The doubled game 384+384=768=7+6+8=__ =2+_=

9 is free from both patterns
9 is free and in NO patterns

9

But if you start doubling 9, it's always 9 !

The 9 doubled game: 9+9=18

The 9 doubled game: 18+18=136

The 9 doubled game: 36+36=7+_
=9

The 9 doubled game: 72+72=144=
1+4+4 =

The 9 doubled game: 144+144=288

The 9 doubled game: 2+8+8=_+_=

9 is free from both patterns

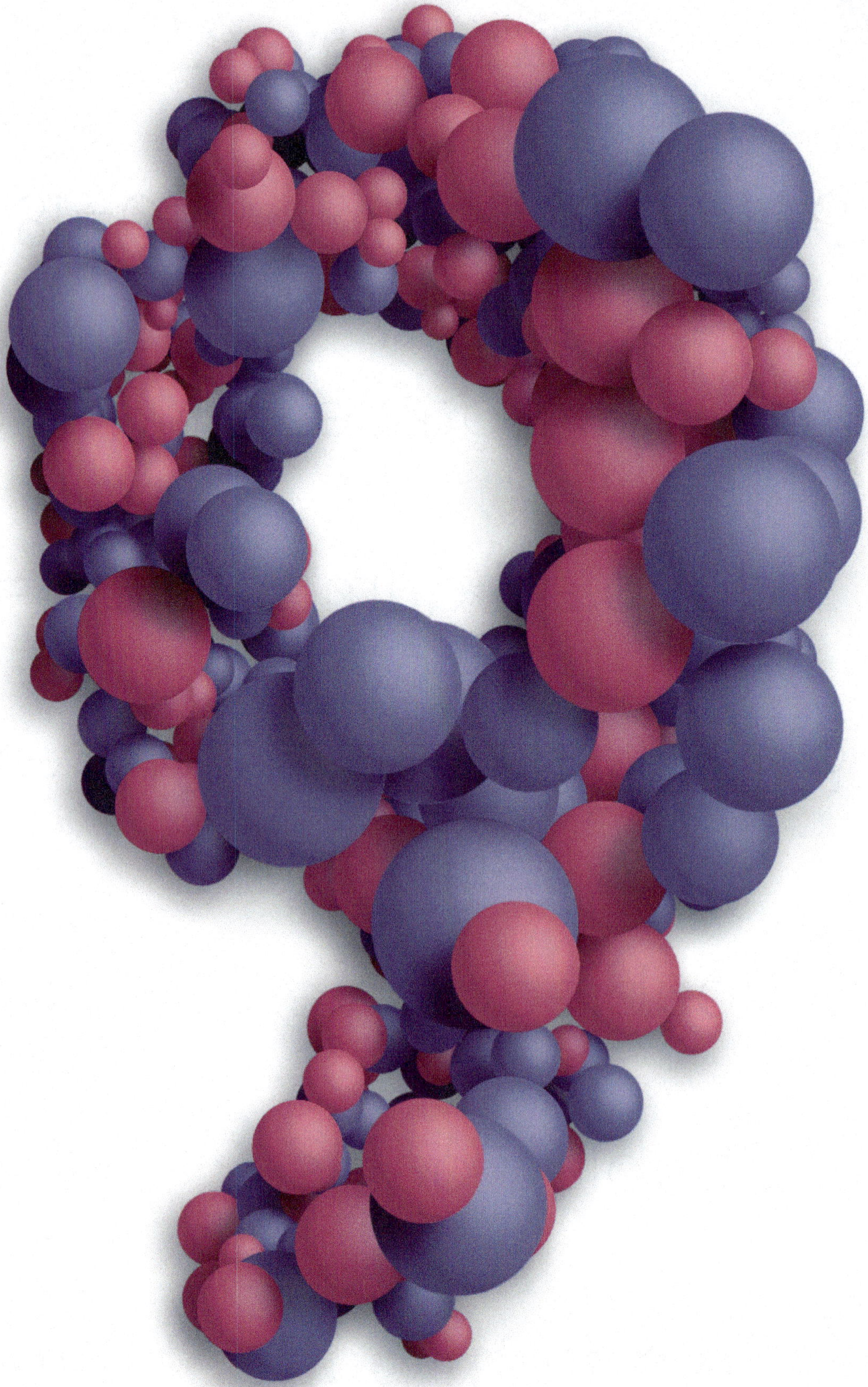

Exercices

But if you start doubling 9, **<u>it's always 9</u>** !

The 9 doubled game: 18+__=36

The 9 doubled game:
36+36=7+_=9

The 9 doubled game: 72+__=144=
1+4+4 =

The 9 doubled game: 144+___=288

The 9 doubled game: 2+8+8=

The 9 doubled game: _+9=18=1+8=

9 is free from both patterns

Exercices

The number 9 is Everywhere in plain sight

The 9 doubled game: 9+9=18=

The 9 doubled game: 18+18=__=

The 9 doubled game:
36+36=7+_=9

The 9 doubled game: 72+72=144=
1+4+4 =

The 9 doubled game: 144+144=288

The 9 doubled game: 2+8+8=

9 doubled is ALWAYS 9

9

Exercices

The number 9 is Everywhere in plain sight

The 9 doubled game: _+9=18

The 9 doubled game: 18+18=1__

The 9 doubled game: 36+36=7+_=9

The 9 doubled game: 72+72=---=
_+_+_ =

The 9 doubled game: 144+144=_+_+_=_

The 9 doubled game: 2+8+8=

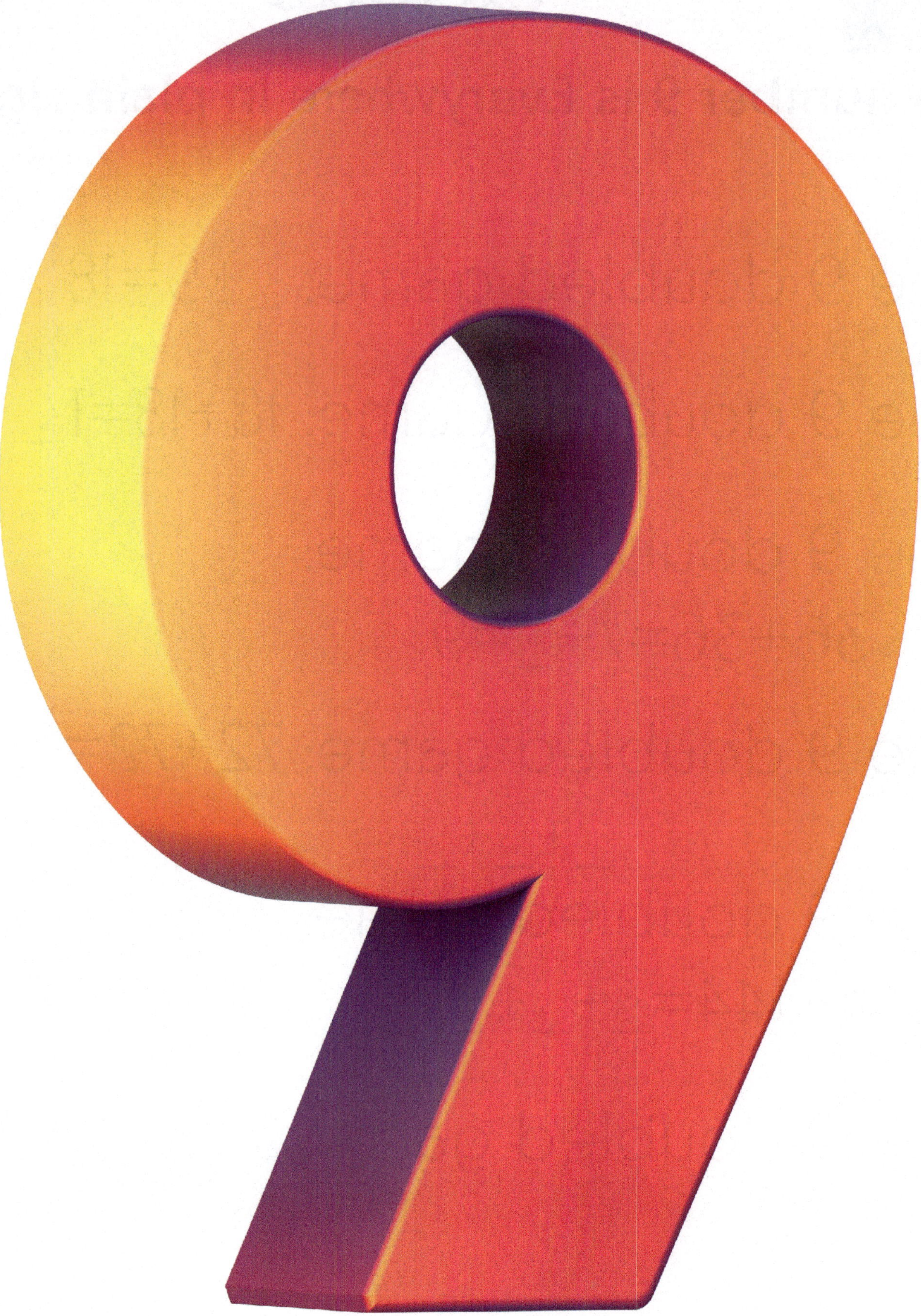
9

Geometric number = 9

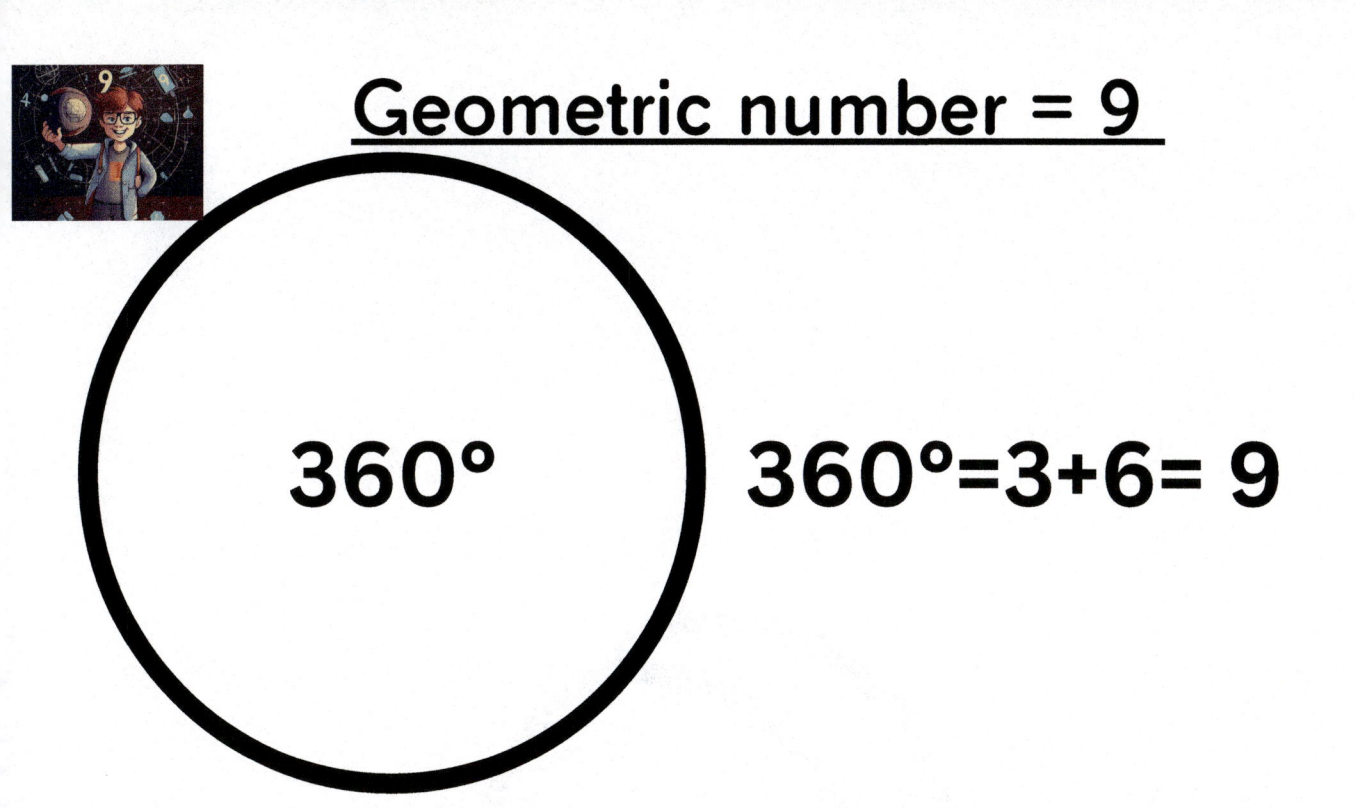

360°

360°=3+6= 9

Geometry measure

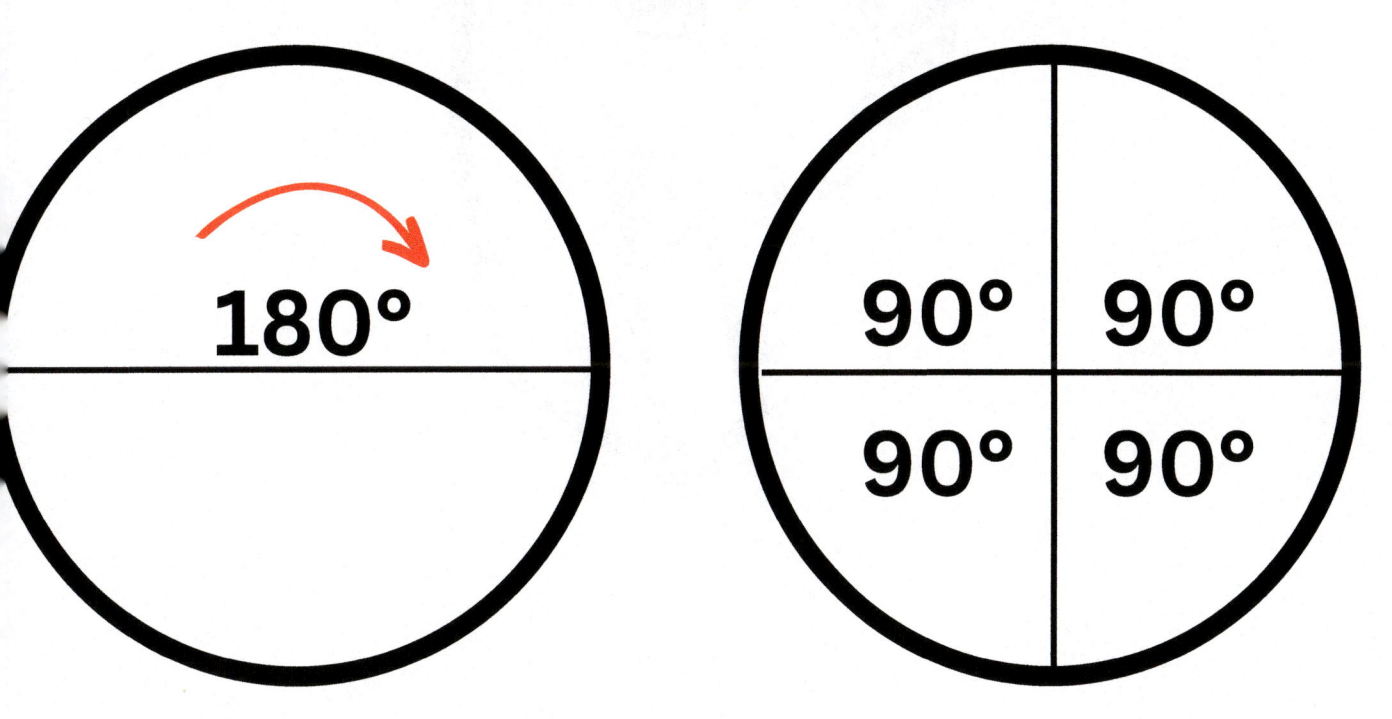

180°

90° | 90°
90° | 90°

180° = 1+8=9

90° = 9+0=9

45° = 4+5= _

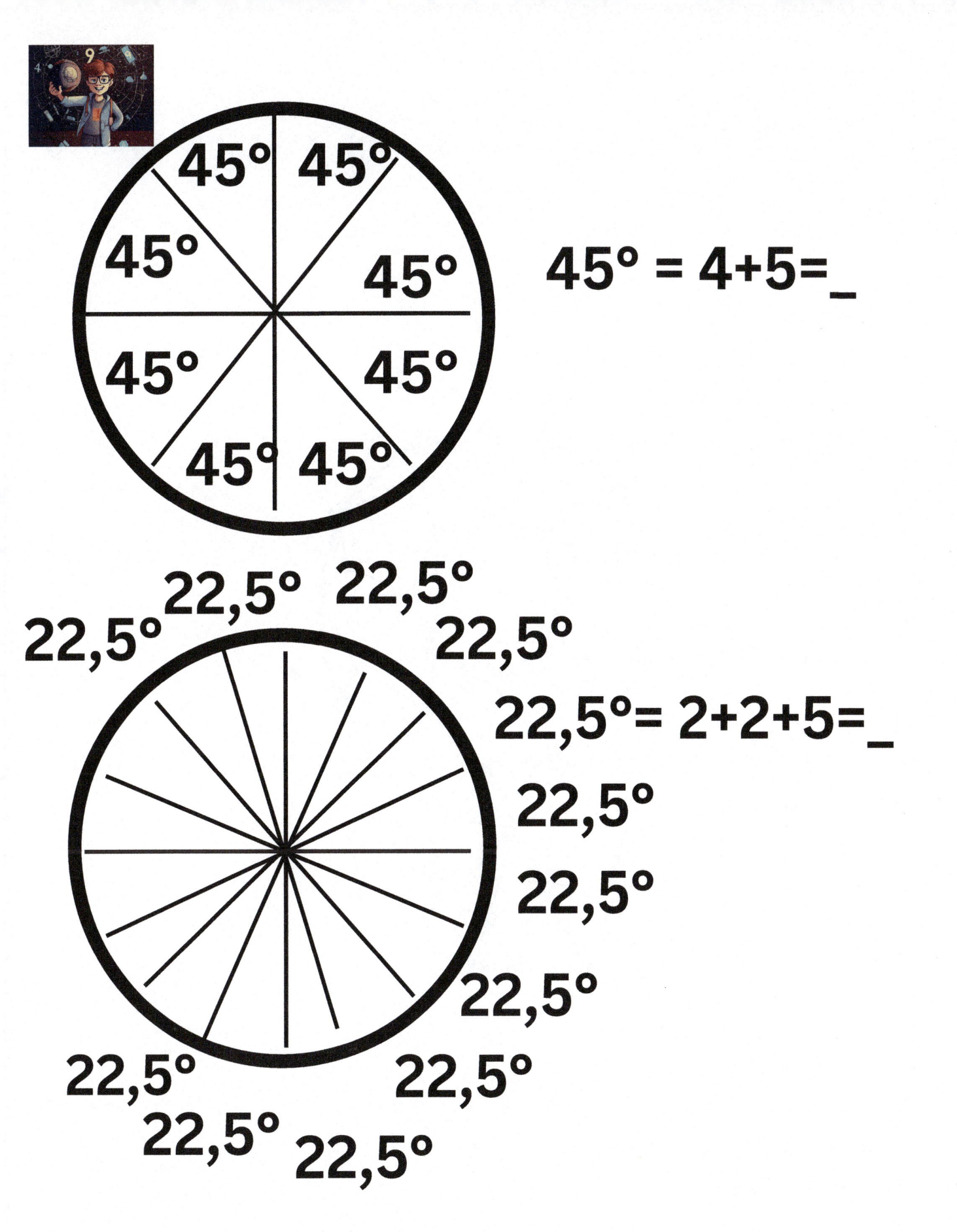

22,5° = 2+2+5= _

Until the infinite...

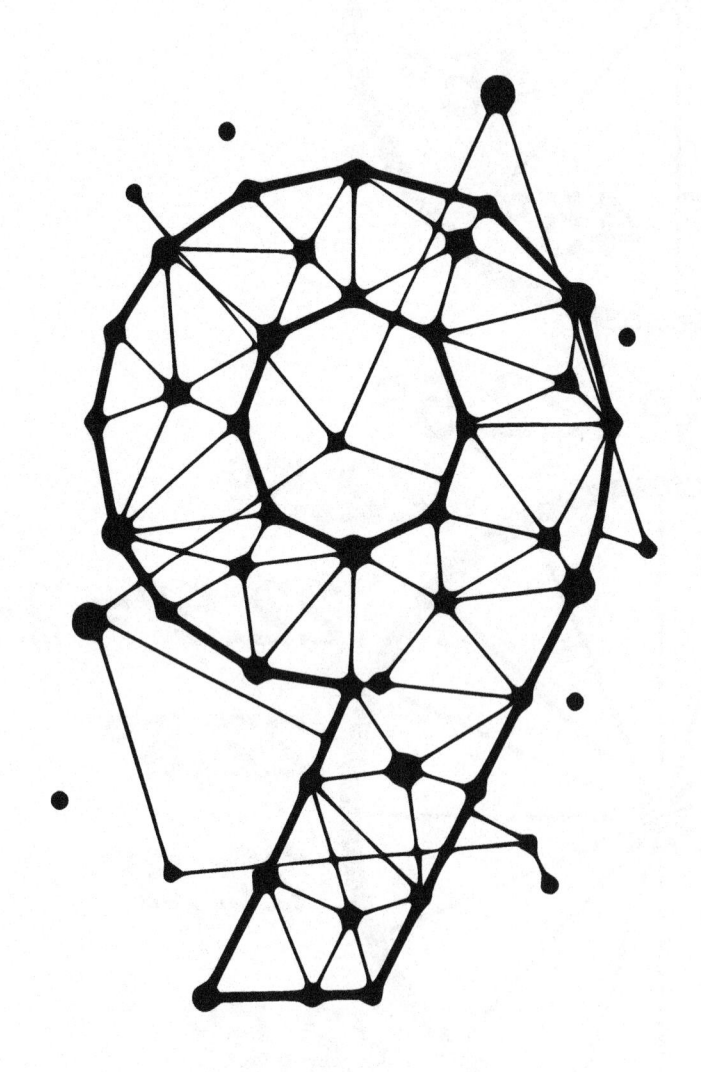

The Alphabet of Sacred Geometry

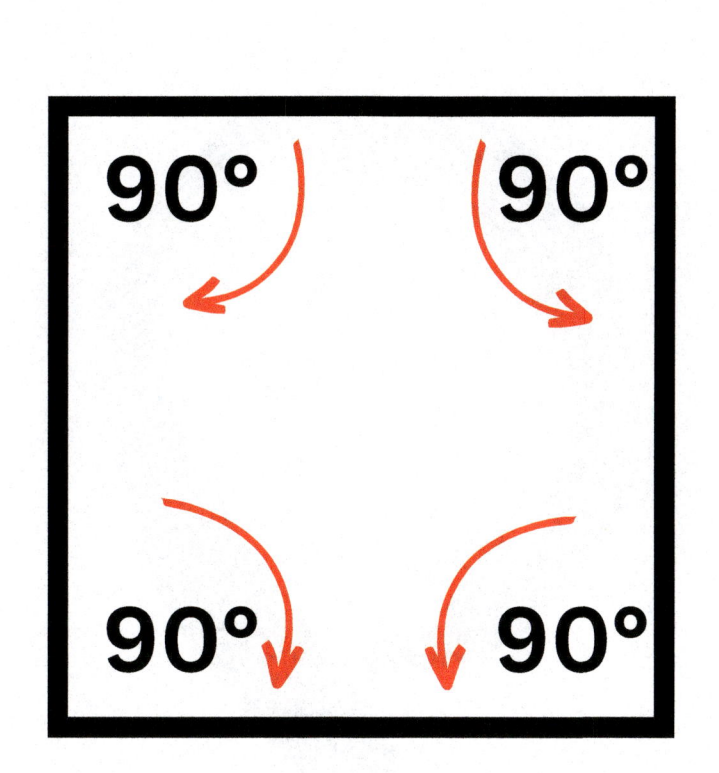

90° × 4 = 360° = 9

Everything ends up to nine !

The Alphabet of Sacred Geometry
Pentagone = 108°

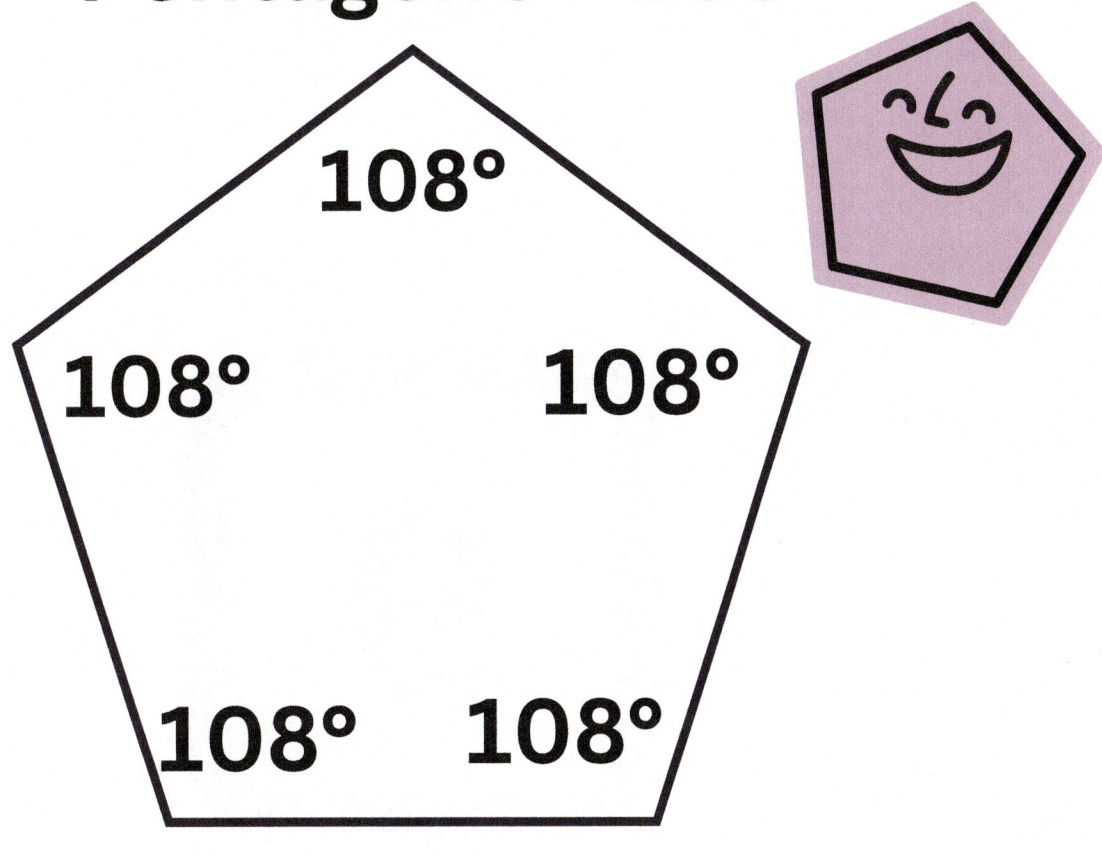

$$108 \times 5 = 540° = 5+4 = 9$$

108 is the silver number in Chemistry, it's 9!

9

The Alphabet of Sacred Geometry
Polygone = 120°

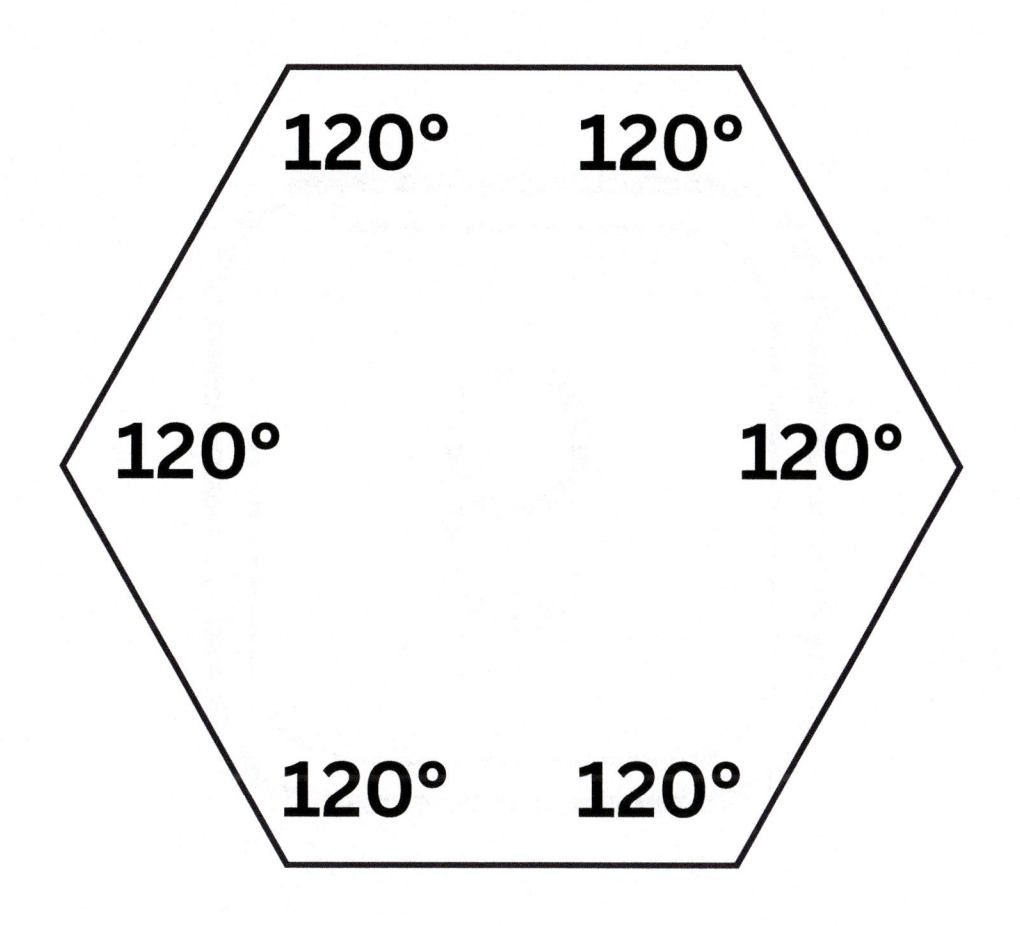

$$120° \times 6 = 720° = 7+2 = 9$$

$$120° \times 6 = 720° = \_+\_=\_$$

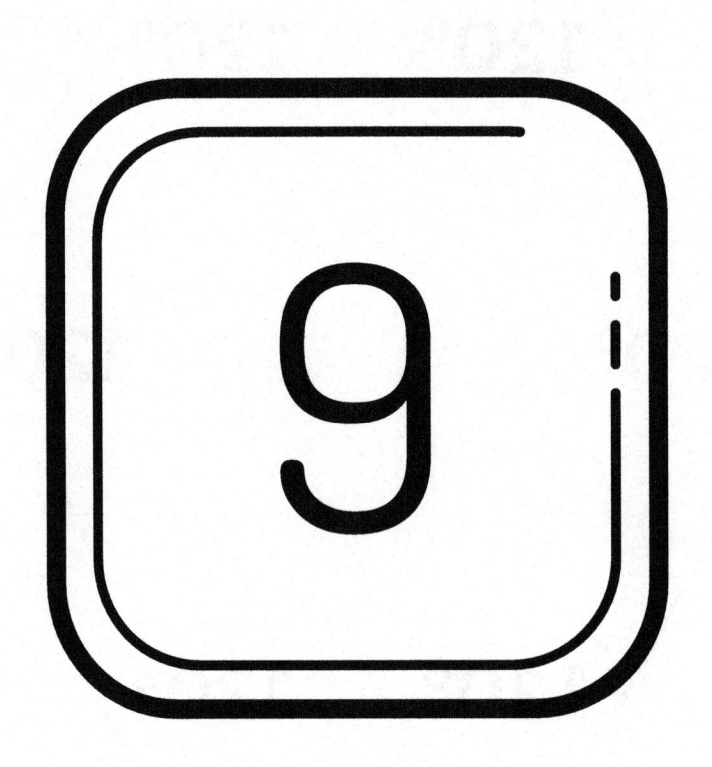

9

The Alphabet of Sacred Geometry

Octogone = 135° = 9

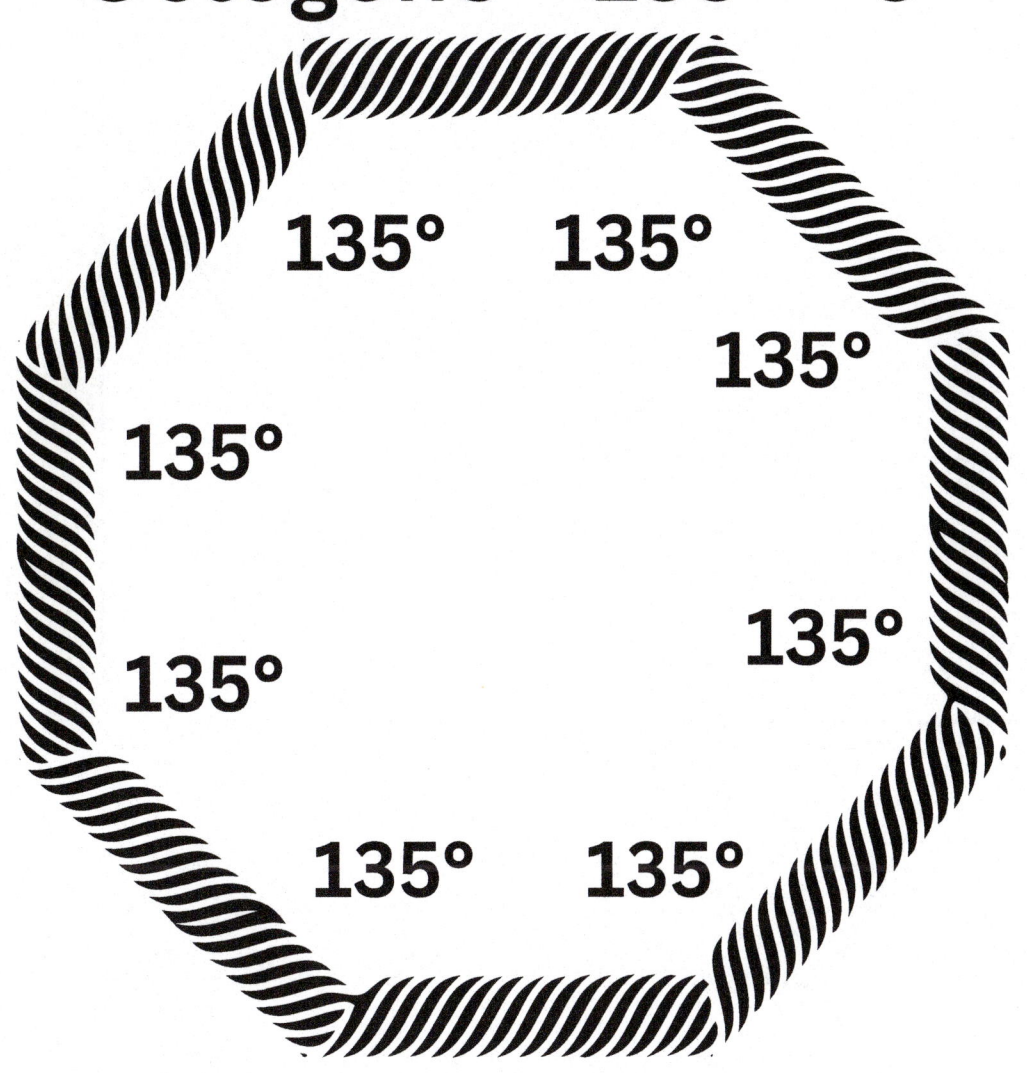

135° 135°
135°
135°
135°
135°
135° 135°

$$135° \times 8 = 1080° = 1+8 = 9$$

TETRAHEDRON = 720°
4 faces x 180° in each vortex

Duration in annual cycle of sacred year

1'440 minutes = (1+4+4=_)

HEXAHEDRON = 2160°
6 faces x 360° in each vortex

cube = 2160°=2+1+6=

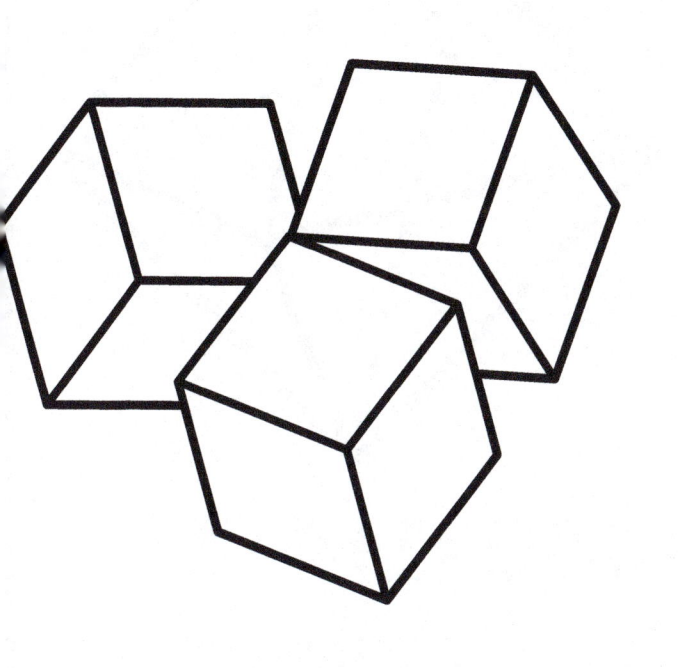

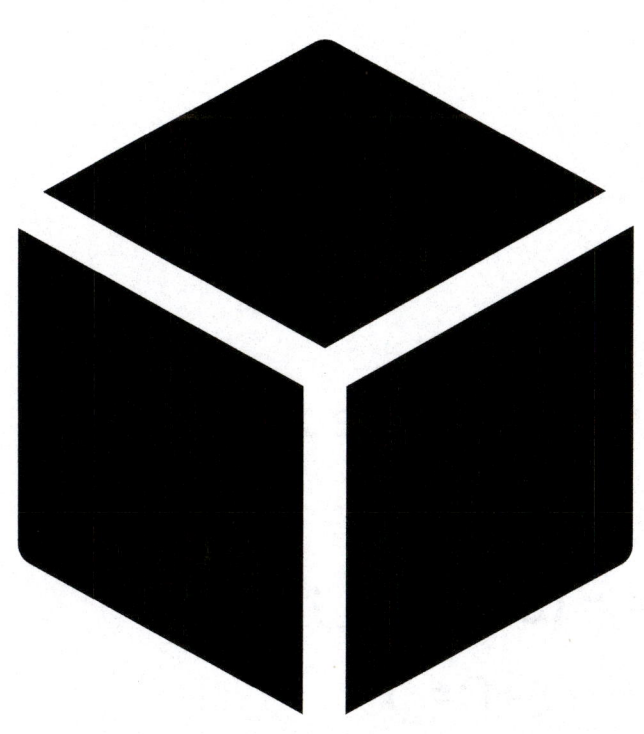

Duration in annual cycle of sacred year

43'200 minutes / 720 hours

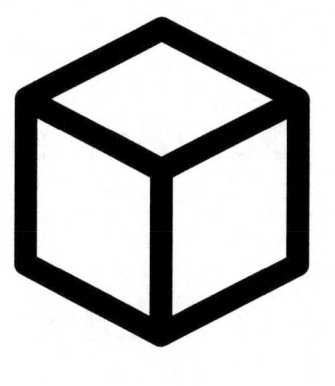

OCTAHEDRON = 1440°
2x3D /2x(4x180) triangle

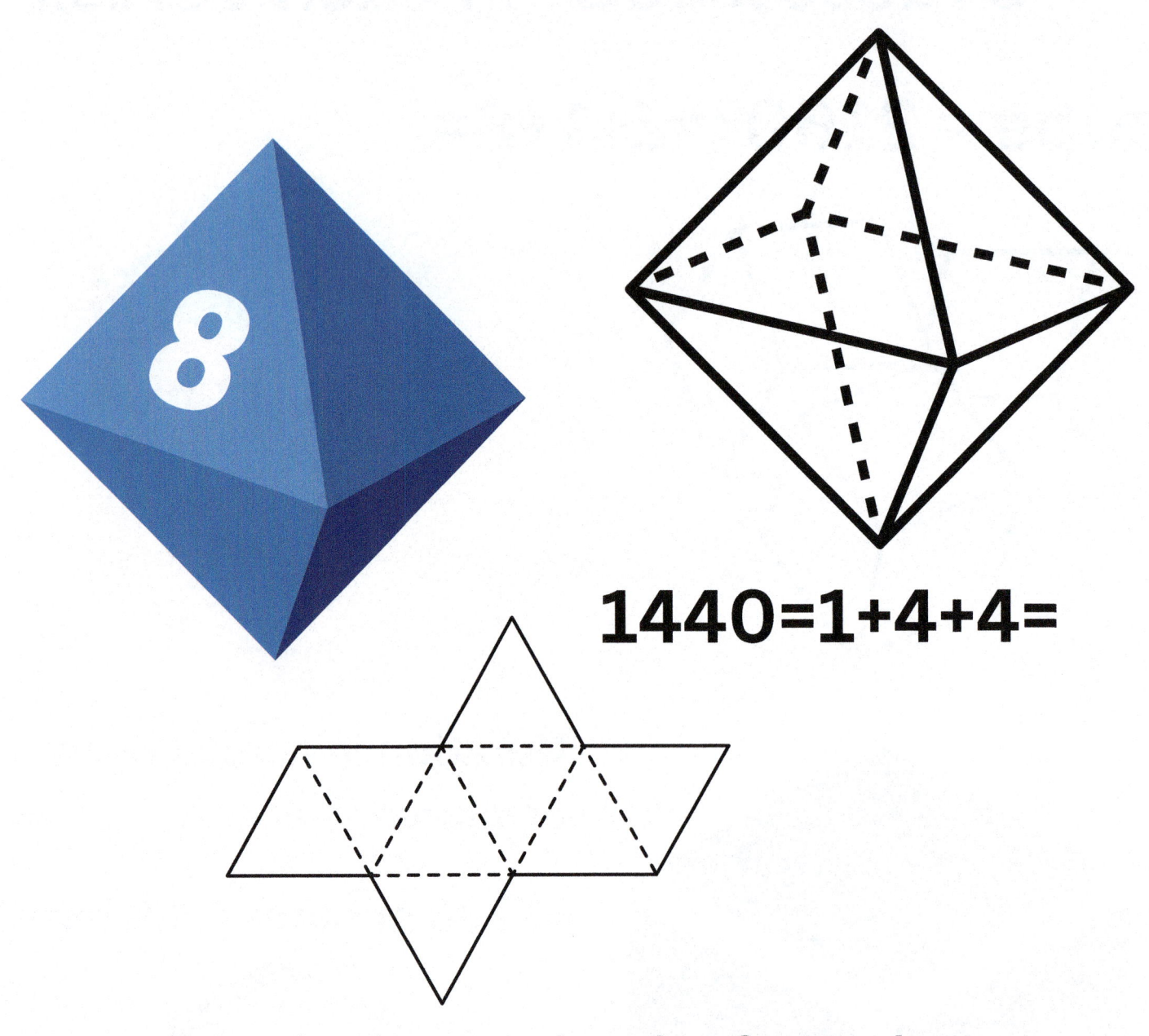

1440=1+4+4=

Duration in annual cycle of sacred year

20 days

28'800 minutes = (2+8+8= _)

480 hours = (4+8+0=_)

ICOSAHEDRON = 3600°
20 faces, 5 faces in each vortex

Duration in annual cycle of sacred year
ICOSAHEDRON = 72'000 minutes

DODECAHEDRON
the shape of the Universe
12 faces
12 x 540° = 6480° =
number 9 again

Duration in annual cycle of sacred year

129'600 minutes / 2160 hours / 90 days

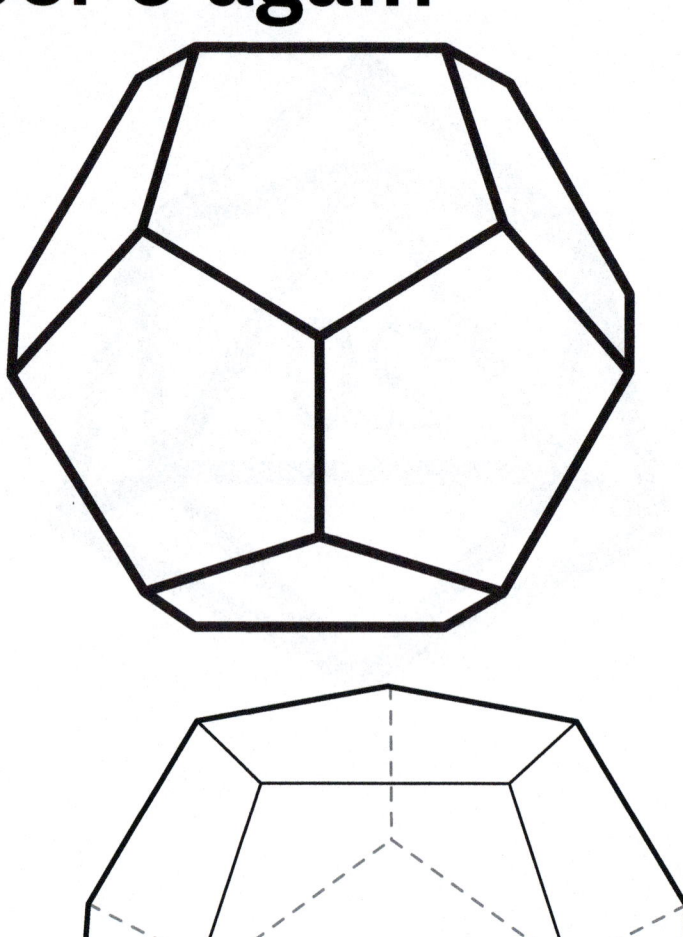

9

Geometry of TIME

**25'920 seconds =
2+5+9+2+0=**

**25'920 hours =
2+5+9+2+0=**

**25'920 days =
2+5+9+2+0=
72 years**

72 years = one life time

TABULATION OF THE VEDIC AGE

NAME OF YUGA

SANDHYA : 36'000

KALI: 360'000

SANDHYAMSA : 36'000

SANDHYA : 72'000

DWAPARA : 720'000

KALI = 360'000 = 9

⑨

9 is everywhere

The Holy Ghost 1080

Earth Orbites around the Sun 66'600 miles/hours

REVELATION 14-1-4= 9

Consciousness Earth 7,83 Hz

Human Design to Cosmis Measures

The Poesy 9 in Music

528 Hz

9 Core Creative Frequencies (not 6)

Everything in the Universe is Made of 9 Notes

6 -> Solfeggio

3 -> Perfect Circle of Sound

528 Hz

9 is a unique number

9 is the number of completion

9 innings to a baseball game

9 Original Knights Templar

9 months of pregnancy

Cool Math Trick Number 9

45 x 45 =

4x5 = 20

5x5 = 25

45 x 45 =

2025 = 2+2+5= 9

Cool Math Trick Number 9

9 always in the middle

8888 x 9 =

9 x 8 = 72

888 = 3 numbers to add , but **9**!

7 _ _ _ 2

9 in the middle

7 **9 9 9** 2

Cool Math Trick Number 9
9 always in the middle

6 x 999 =

6 x 999 = 5 _ _ 4

6 x 9**99** = 5 **9 9** 4

7 x 9999 =

7 x 9999 =

7 x 9999 = 6 _ _ _ 3

7 x 9**999** = 6 **9 9 9** 3

Cool Math Trick Number 9
above 9 minus 1 and minus the number 1000 here 11

$$11 \times 999 =$$

$$11 \times 999 = 10$$

$$11 \times 999 = 10 - \quad 1000-11=989$$
$$-1$$

$$10'989$$

$$12 \times 999 =$$

$$1000-12=988$$
$$12 \times 999 = 11\;988$$
$$-1$$

Cool Math Trick Number 9

14
-9

19
-9

1+4 =5
-9

5

1+9 =10
-9

10

9 and Fibonacci

Position | 1 | 2 | 3 | 4 | 5 | 6 | 7 | 8 | 9 | 10 | 11 | 12

Fibonacci | 1 | 1 | 2 | 3 | 5 | 8 | 13 | 21 | 34 | 55 | 89 | 144

12 positions later | 233 | 377 | 610 | 987 | 1597 | 2584 | 4181 | 6765 | 10946 | 17711 | 28657 | 46368

Sum of digits | 8 | 17 | 7 | 24 | 22 | 19 | 15 | 18 | 18 | 16 | 28 | 27

Reduced | 8 | 8 | 7 | 6 | 4 | 1 | 6 | 9 | 9 | 7 | 10 (1+0) | 9

9 | 9 | 9 | 9 | 9 | 9 | 9 | 9 | 9 | 9 | 9 | 9 | 9

**The pattern is striking: when we add the Fibonacci number to the reduced sum of the digits of the number 12 positions later,
we consistently get the number 9.**

9

Printed in Great Britain
by Amazon